SAMSUNG GALAXY A05 GUIDEBOOK FOR BEGINNERS AND SENIORS

A COMPLETE GUIDE WITH THE LATEST ANDROID TIPS & TRICKS ON HOW USE THE NEW SAMSUNG GALAXY A05 SMARTPHONE LIKE A PRO.

DAVIS MOORE

Table of Contents

Chapter One 14

Device Layout and Functions 14

Chapter Two 15

Hard Keys 15

    Setting for the Side key (Power button) 16

    Soft buttons 16

Charging the Device Battery 17

    Wired charging 17

    Decreasing the battery consumption 18

    Tips and precautions for charging the device Battery 19

SIM or USIM card (Nano-SIM card) 20

    Installing the SIM or USIM card 20

    SIM card manager (Dual SIM models) 21

Micro SD card 22

    Insert a memory card 22

    Take out the memory card 24

    Formatting the memory card 25

Switch the Device on and off 26

    Power on the device 26

Power off the device 26

Force reboot 27

Emergency mode 27

Initial setup 28

Samsung account 28

Finding your ID and resetting your password 29

Signing out of your Samsung account 29

How to use Smart Switch to importing data from your old phone 30

Transferring data wirelessly 30

Restore and back up data using external storage 31

Import backup data from a computer 31

How to control the touch screen 33

Navigation bar (soft buttons) 34

Hiding the navigation bar 35

Chapter Three 36

Home and Apps screen 36

Switching between Home and Apps screens 36

Editing the Home screen 37

Revealing all apps on the Home screen 39

Launching Finder 39

- Moving items 39
- Creating folders 40
- Edge panel 42
- Lock screen 43
  - How to change the lock screen method 43
- Indicator icons 44
- Screen capture 45
- Notification panel 47
  - Making use of quick setting buttons 47
  - Controlling media playback 48
  - Controlling nearby devices 49
- Inputting text 50
  - Keyboard layout 50
  - How to change the input language 50
  - How to change the keyboard 50
  - Additional keyboard functions 51
  - Copying and pasting 52
- Application and features 53
  - Galaxy Store 53
  - Play Store 53
  - Uninstalling or disabling apps 54

Enabling applications 54
Setting app permissions 54
Chapter Four 56
Phone application 56
Making calls 56
Making calls from call history or contacts list 57
Making an international call 57
Answering a call 57
Rejecting a call 57
Blocking phone numbers 58
Options during calls 59
Contacts 60
Creating a new contact 60
Importing contacts 60
Syncing contacts with your web accounts 60
Searching for contacts 61
Deleting contacts 61
Sharing contacts 62
Creating groups 62
Merging duplicate contacts 62
Chapter Five 63

Messages application 63
    Sending messages 63
    Viewing messages 64
    Sorting messages 65
    Deleting messages 65
    Changing message settings 65
Internet 66
    Using secret mode 67
Chapter Six 68
Camera 68
    Camera etiquette 68
    Taking pictures 68
    Using zoom features 70
    Locking the focus (AF) and exposure (AE) 71
    Using the camera button 71
    Options for current shooting mode 72
    Photo mode 73
    Taking selfies 73
    Applying filter and beauty effects 74
    Video mode 74
    Portrait mode 75

    Pro mode 76

    Panorama mode 76

    Food mode 77

    Macro mode 78

    Hyperlapse mode 78

    Deco Pic mode 78

Gallery 79

    Using Gallery 79

    Grouping alike images 80

    Viewing images 80

    Cropping enlarged images 81

    Viewing videos 81

    Albums 82

    Stories 83

    Using the Trash feature 83

Multi window 84

    Split screen view 84

    Launching apps from the Edge panel 85

    Adding app pairs 86

    Adjusting the window size 86

    Pop-up view 86

Launching apps from the Edge panel 87
Samsung Notes 88
    Creating notes 88
    Deleting notes 89
Galaxy Shop 89
Calendar 89
    Creating events 89
    Syncing events with your accounts 89
Reminder 90
    Starting Reminder 90
    Creating reminders 90
    Finishing reminders 91
    Restoring reminders 91
    Deleting reminders 91
Radio 92
    Playing through the speaker 92
My Files 93
Clock 93
Calculator 94
Sharing content 95
    Quickly Share content with nearby devices 95

Google apps 96

Chapter Seven 98

Settings 98

Samsung account 98

Connections 98

    Options 98

Wi-Fi 100

    Connecting to a Wi-Fi network 100

    Wi-Fi Direct 101

Bluetooth 102

    Pairing with other Bluetooth devices 102

    Sending and receiving data 103

Data saver 104

    Mobile data only apps 105

Mobile Hotspot 106

Printing 107

    Adding printer plug-ins 107

    Printing document 108

Sounds and vibration 109

    Options 109

Notifications 111

- Display 112
  - Wallpaper 114
  - Themes 114
  - Home screen 114
  - Lock screen 114
  - Smart Lock 116
- Chapter Eight 117
- Biometrics and security 117
  - Face recognition 119
  - For better face recognition 119
  - Registering your face 120
  - Unlocking the screen with your face 120
  - Deleting the registered face data 121
  - Fingerprint recognition 121
  - For better fingerprint recognition 122
  - Registering fingerprints 123
  - Unlocking the screen with your fingerprints 124
  - Deleting registered fingerprints 124

Introduction

As a beginner or senior user, this book will enhance your smartphone knowledge and also help you get the most out of your device.

This book is a comprehensive step by step guide with photographs that will enable you to easily operating or find settings on the smartphone and it also contains numerous advanced features of the Galaxy A05 that are rarely seen elsewhere.

It's satisfactory for all level of beginner and senior.

Copyright 2024 © Davis Moore

All rights reserved. This book is copyrighted and no part of it may be reproduced, stored, or transmitted, in any form or means, without the prior written permission of the copyright owner. Printed in the United States of America.

Copyright 2024 © Davis Moore

Chapter One
Device Layout and Functions

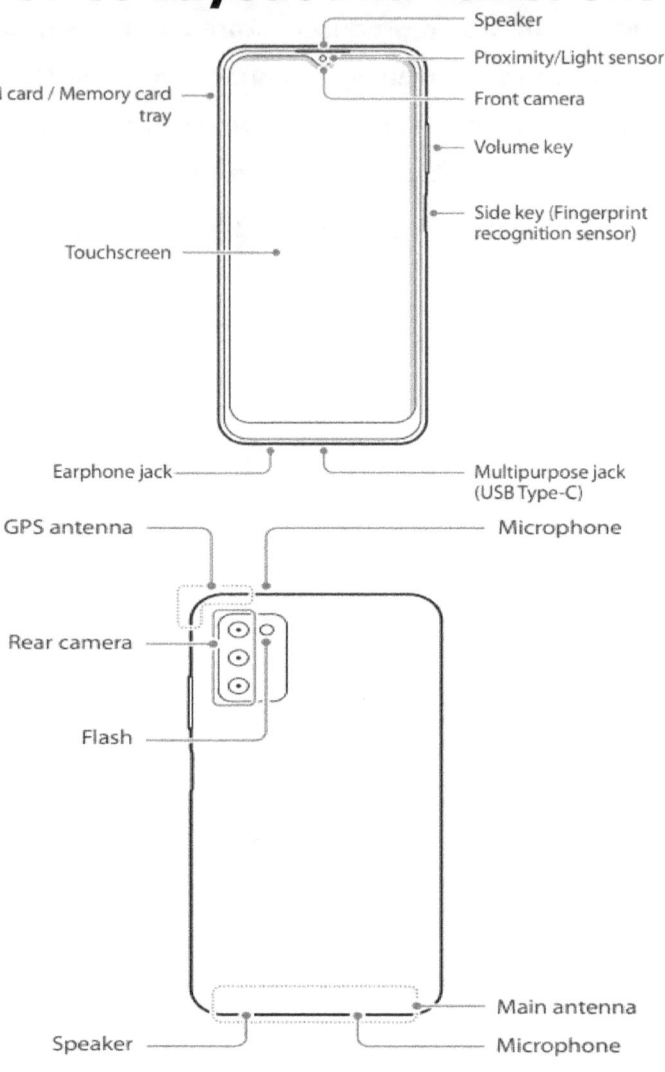

Chapter Two
Hard Keys

KEY	FUNCTION
Side Key	- Press and hold the side button to turn on or off the smartphone. - Tap the Side button once to lock or unlock the device. - Double press the side key to open applications that have been set to open. - Place your finger on the Side button to enroll your fingerprint.
Side Key + Volume Down Key	- Pressing the Side key and volume down button allows you to take a screenshot. - Hold these keys down for some time to turn off the device.

Setting for the Side key (Power button)

Apps or features can be launched by pressing Side key twice.

- o Launch the Settings application and select the Advanced features menu then click Side key. Select a preferred option.

Soft buttons

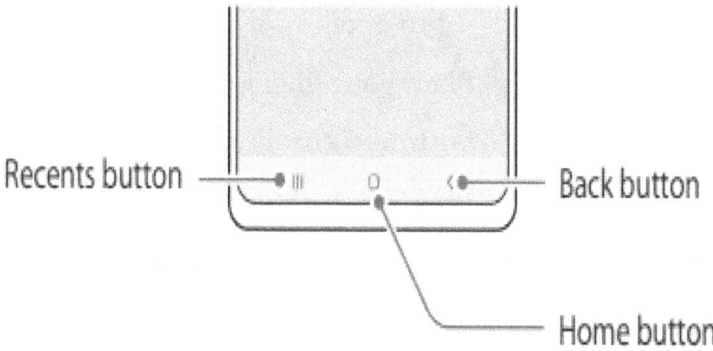

When your gadget is activated, the soft buttons will appear on the bottom of the screen, these buttons contains:

- Home button
- Back button
- Recent button

Charging the Device Battery

Your gadget should be charged before turning it on for the first time. Do the same if the gadget hasn't been used for a long time.

Caution: Your gadget will face damages if the original Samsung charger that was allocated to the device was not used.

Precaution:

- If the charger is connected wrongly, it will cause damage to the gadget.

    Note: Any damage that is caused by this is not covered by warranty.

- Only Samsung Type-C cord should be used. Using others may harm the gadget.

Wired charging

Connect the USB Type-C cord to the power adapter then connect the other end of the cord to the multipurpose jack of the gadget. Make sure that you disconnect the charger from the gadget when it is fully charged.

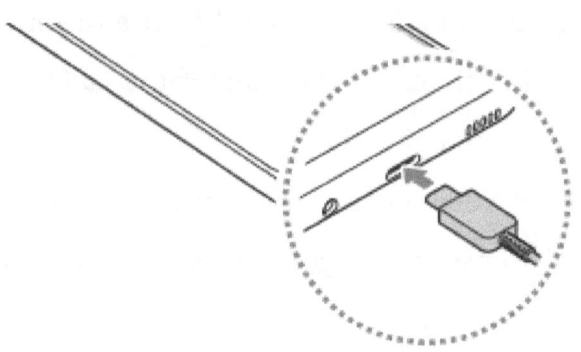

Decreasing the battery consumption

There are various options of this device that will help conserve your battery.

- Make use of the care feature to enhance the phone.
- Turn off the screen always when it is idle.
- Activate the Power saving mode feature.
- Closed applications that are not in use.
- Turn off Bluetooth when it is not in use.
- Decrease the backlight time.
- Decrease the brightness of the screen.

Tips and precautions for charging the device Battery

- When network apps or multiple apps are in use, battery drains faster. Charge gadget fully before transferring files to avoid losing power.
- Other charging means except the prescribed process may cause damage to the gadget.
- Charging speed will decrease when you are using it while charging.
- There will be an unstable function of the screen if there is an unstable power supply during device charge. If this occurs, unplug the phone from the charger.
- If the multipurpose jacket of the device is wet, do not charge. Allow it to dry up first.
- See the Samsung Service Center for more information.

SIM or USIM card (Nano-SIM card)

Insert a SIM card or a USIM card that is provided by your device service provider.

The speed of data transfer when using dual SIM card may sometime not be compared to that of a single SIM card.

Installing the SIM or USIM card

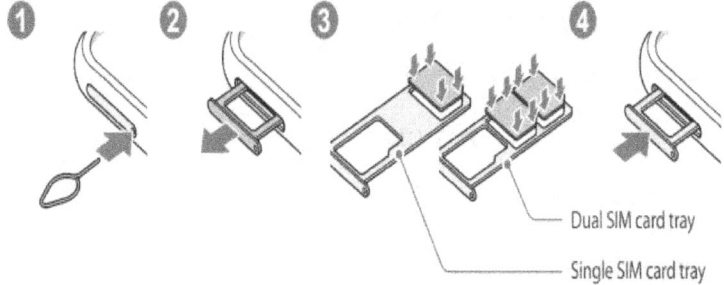

1. Insert the pin of ejection in the hole to bring the SIM card tray out.
2. Take out the slot carefully.
3. Insert SIM card with the part colored gold facing downward.
4. Push the tray into the slot after putting the SIM cards.

Precautions:

- Only a Nano-SIM card should be used.
- The gadget may be injured if the pin of injection is not in the right angle.
- If you don't fix the SIM properly, it may fall out.
- Allow the tray to dry up first before putting it into your gadget to avoid damage.
- Make sure that you insert the tray into the hole to avoid liquid from entering your device.

SIM card manager (Dual SIM models)

Go to Settings and tap Connections and click SIM card manager.

- **SIM cards**: Select a SIM card to use and change its settings.
- **Preferred SIM card**: Choose a SIM to utilize for calls and messages, if you've two SIMS on your device.
- **More SIM card settings**: To change the settings for calls, select More SIM card settings.

Micro SD card

Insert a memory card

Use the images below to install a memory card. Depending on your card type or manufacturer of the gadget, some memory cards may not be suitable for the use of your device.

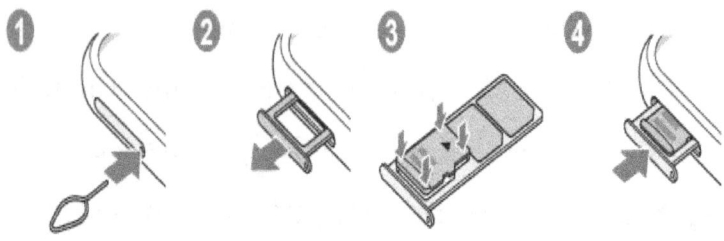

1. To take out the memory card tray, put the ejection pin into the hole.
2. Take the tray out carefully.
3. Place it with gold colored part facing downwards.
4. Place the tray back into the slot after inserting the memory card.

Precaution:

- Just know that not all memory cards can be used in your gadget. All data stored in a mismatch memory card may get corrupt.

- Make sure that you rightfully insert the memory card.
- Ensure the pin is in the right angle to avoid injury of the gadget.
- When the tray is taken out of the device, mobile data will turn off automatically.
- Ensure tight placement of the memory card to avoid it falling off.
- Make sure that the tray is dried if wet before inserting the memory card.
- Insert the memory card into the tray and the tray into the device properly to prevent liquid from entering it.

Note:
- This gadget supports FAT exFAT file systems for memory cards. If a memory card that is formatted in another system is put into the device, the device will not detect the card or ask you to reformat it. Go to a Samsung Service Center if you are unable to format the memory card with your gadget.

- Frequent deletion of data and media should not be done on the memory card if you want it to last longer.
- An instruction will appear in the screen and shoe My Files then click SD card folder, if a memory card is out into the gadget.

Take out the memory card

If you don't unmount the memory card, please do not remove it.

1. To unmount the SD card, go to the Settings app and select Battery and device care and go to the ⋮More options menu
2. Click Advanced and tap SD card the click on Unmount.

Precaution: Ensure not to take away the memory card if you are transferring data or accessing information. If this is done file might get corrupt and the memory card might also get spoilt.

Formatting the memory card

You may not be able to use the memory card on your gadget if it is formatted on a computer. Make sure you format it on your device.

1. To format a memory card, go to the Settings application and click on Battery and device care the tap Storage
2. Select:More options and tap Advanced then SD card and finally tap Format.

Precaution: Always backup important data saved in the memory card before formatting it. And take note that the manufacturer's warranty does not cover data loss caused by your misuse.

Switch the Device on and off

Note: Follow guideline and laid down rule while in the hospital or an airplane to know where, when and how to use your device.

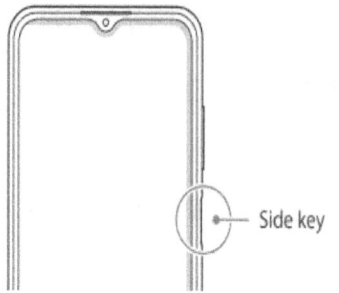

Power on the device
Press and hold the Side key for a while to power on the device.

Power off the device
1. Press and hold the side key to power the gadget off. Go to the notification panel and touch ⏻ to power off the device.

    Note: To access the notification panel from any part of your screen, swipe downwards from the top of the screen.

2. Tap Power off the turn the gadget off.

    To put the gadget off and on again, tap Restart.

Force reboot

Forcefully restart your gadget if it becomes unresponsive. Press and hold the Side key and Volume downward key for 10 seconds to restart the device forcefully.

Emergency mode

With the emergency mode turned on, your gadget's battery lasts longer and some apps will be restricted. With this feature, you can call emergency numbers and send your current location to friends.

To turn this feature on, press and hold the Side key then select Emergency mode. Go to the notification panel and select ⏻ then click Emergency to turn it on also.

To disable this mode on your gadget, tap Turn off emergency mode under the ⋮ More options menu.

Initial setup

To finish the setup of your device after turning it on for the first time after purchasing it, follow the commands on the screen.

Note: Some feature can be set without a Wi-Fi network connection, during the setup while some requires Wi-Fi connection.

Samsung account

Having a Samsung account grants you access to some exclusive features of Samsung, and allows you to make use of Samsung services like TVs, mobile device and the official Samsung website.

To get acquainted with the list of services that can be used with a Samsung account, visit account.samsung.com.

1. To create an account, go to the Settings menu and click on Account and backup the Manage account then Add account and finally click on Samsung account.
2. If you have an account already, enter it or create another if you don't have.

- Choose Continue with Google to sign in with Google account.
- Choose Create account to generate a new Samsung account.

Finding your ID and resetting your password

On the sign in screen, click Find ID or Reset password to retrieve your account password if you don't remember.

Signing out of your Samsung account

When an account is logged out, events and contacts on the account will leave the device.

1. From Settings, click on Account and backup and choose Manage accounts.
2. Touch Sign out under the screen to log out your account after tapping the Samsung account and click My Profile.
3. After clicking Sign out, enter your password and click OK.

How to use Smart Switch to importing data from your old phone

The Smart Switch is an easy and convenient way to transfer data and media from an old device to the new one. Launch Settings and tap Account and backup then Bring data from old device.

Note: This feature cannot work on unsupported devices.

Transferring data wirelessly

Do the following to perform a wireless data transfers

1. Go to the Smart Switch application on the old device. If you don't have the app, go and get it from the Galaxy Store or Google Play Store.
2. Go to the Settings app and click on Account and backup the select Bring data from old device.
3. Bring the two devices together.
4. Choose Send data and select Wireless on the new device.
5. Choose allow in the old phone.
6. Select items and select transfer.

Restore and back up data using external storage

Backup all your data in an SD card.

1. Your device media can be backed up to a flash drive, SD card or any other external storage device.
2. Connect the external storage device to your gadget.
3. Go to the Setting application and click on Account and backup the select the External storage transfer option.
4. Select Restore from SD card and tap Restore.
5. Tag along the rules on the screen to transfer data from an external storage.

Import backup data from a computer

Go to www.samsung.com/smartswitch to install the computer version of the Smart Switch application to transfer data smoothly from a computer to a device. Your data on your old phone should be backed up on a PC then import it to your new device.

1. Visit the www.samsung.com/smartswitch website to download Smart Switch on your PC.

2. Launch the software on your computer.

   Note: If the previous device is not a Samsung product, transfer and backup data to the computer using the software allocated by the device manufacturer.

4. Use a USB cord to join the device to a computer.

5. Make sure that you follow the instructions on the computer screen to transfer and backup data.

   Note: After transferring and backing up data, unplug the phone from the computer.

6. Connect your new device to the computer using a USB cable also.

How to control the touch screen

Tapping
Tap the screen.

Tapping and holding
Tap and hold the screen for approximately 2 seconds.

Dragging
Tap and hold an item and drag it to the target position.

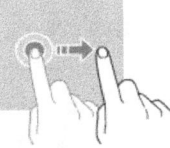

Double-tapping
Double-tap the screen.

Swiping
Swipe upwards, downwards, to the left, or to the right.

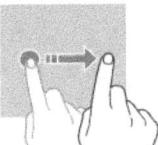

Spreading and pinching
Spread two fingers apart or pinch on the screen.

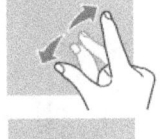

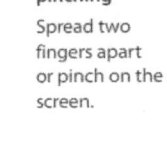

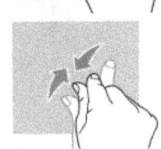

Precaution:

- Do not bring your phone's screen close to any electrical device. When electrostatic discharges take place, the device screen may not function well.

- Don't touch the screen with object that are metallic in nature to avoid scratches that may damage the screen.

Note: Touch inputs that are made to the edge of the touch screen may not be recognized.

Navigation bar (soft buttons)

Your navigation bar is located at the bottom of your screen, and the soft buttons will appear on it when the screen of the phone is turned on.

BUTTONS	FUNCTION
⦀ Recent	Tap this button to go to the apps that have recently been used.
◯ Home	Click this button once to go back to the Home screen. Touch and hold the button to launch Google Assistant.
＜ Back	Tap to return to the previous screen that was just minimized.

Hiding the navigation bar

If you want to enjoy a wider screen, hide the navigation button on the device.

- From Settings, click Display and tap Navigation bar then select Swipe gesture and finally tap navigation type. When the navigation button is hidden, navigation hint will display. To choose an option, click, More options.

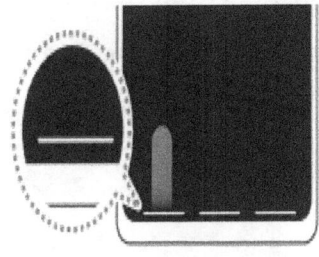

Swipe from bottom

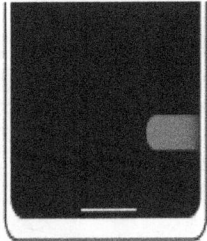

Swipe from sides and bottom

Gesture hint can also be hidden at the bottom of the screen if you don't want it. To hide it tap the switch for Gesture hints.

Chapter Three
Home and Apps screen

All the features of this device are available on the Home screen.

All application icons appear on the Apps screen.

Switching between Home and Apps screens

To open the Apps panel, swipe upwards from the bottom of your device while on the Home screen.

Swipe downwards, to go back to the Home screen from the Apps screen, you can also go back to the Home screen by using the Back button.

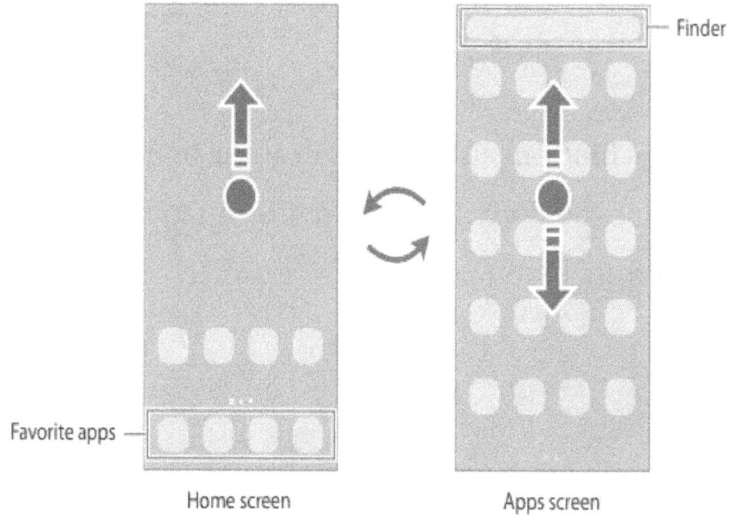

Add the Apps button to the Home screen to access it easily.

Tip: Press and hold and empty space in the Home screen and click Settings then tap Show Apps screen button on Home screen to activate this feature. The app button will appear as show below.

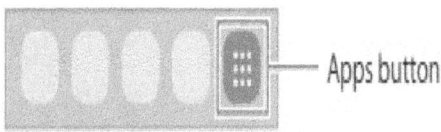

Editing the Home screen

Press and hold an empty space on the Home screen to view thee editing options. Wallpapers and widgets can be added or deleted from the Home screen.

- Swipe the tip of your finger to the left, and then click the ⊕ Plus icon to add panels.

- Press and hold a panel preview then move it to a new location to change it place.

- Select 🗑 on the panel to delete the panel.

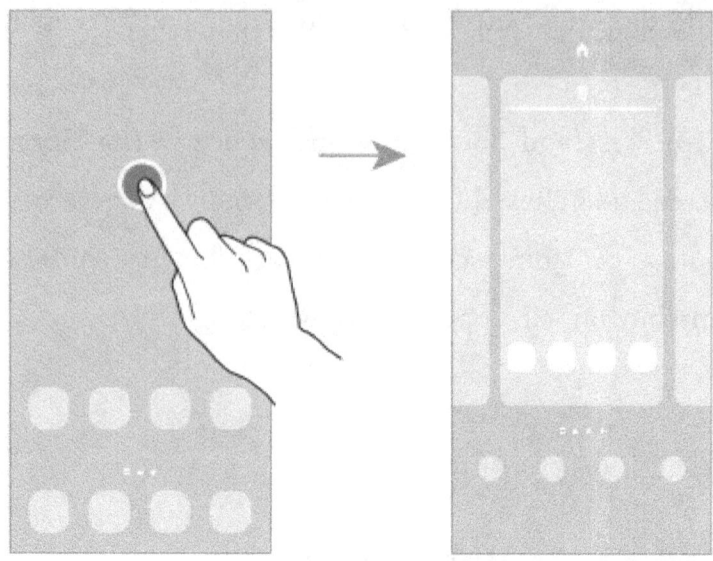

- **Wallpapers**: The wallpapers setting of your device Home and Lock screen can be changed.

- **Themes**: Once a theme is selected, your icons, colors and wallpapers will change automatically.

- **Widgets**: To add a widget to your Home screen, press and hold and empty space, then choose the widget and select Add.

- **Settings**: From your Home screen the screen layout can be changed.

Revealing all apps on the Home screen

Configure your gadget to display all the application of the device on the Home screen:

To do this, enter the Home screen and Tap Settings then click on Home screen layout and tap Home Screen only and choose apply.

Launching Finder

With the feature you can look for apps on your device with ease.

1. Go to the App screen and tap Search or tap Q on the Notification panel.
2. Input your search criteria and matching results will appear on the screen.

   More content can be looked up for when you tap Q on the keyboard.

Moving items

Press and hold an app or item that you want to move to a new location then move it to where you want it to be.

If you want the App shortcut to be added to your Home screen, touch and hold the App icon on the

Apps screen and select the Add to Home option. Apps that are used often can be added to your shortcuts are on the lower area of your Home screen.

Creating folders

To create a folder on the Home Screen or Apps screen, press an app icon and move it over another.

A folder which is containing selected apps will be created automatically. Click Folder name to Name your folder.

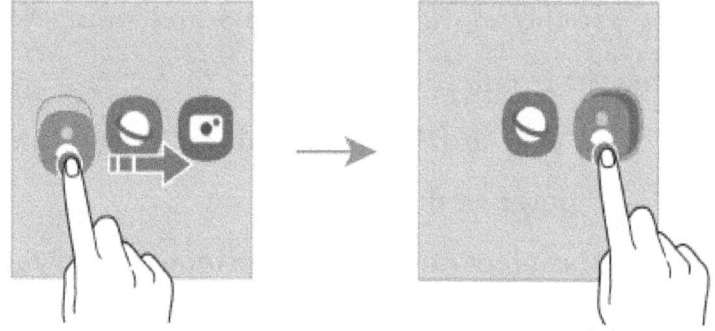

- **Adding more apps**

    To add apps to the folder you have created, tap ➕ and select the apps then click Done.

- **Moving apps from a folder**

    Touch and hold an app to take it to a new location.

- **Deleting a folder**

    Touch and hold the folder that you wish to delete and tap Delete folder. The apps in the folder are not deleted but taken into the Apps screen.

Edge panel

Often used apps can be accessed easily when they are added to the Edge panel.

Take the button customized for the Edge panel to the edge of the screen.

Move to settings directly and click on Display the Edge panel setting switch to activate the button if it does not appear on the screen.

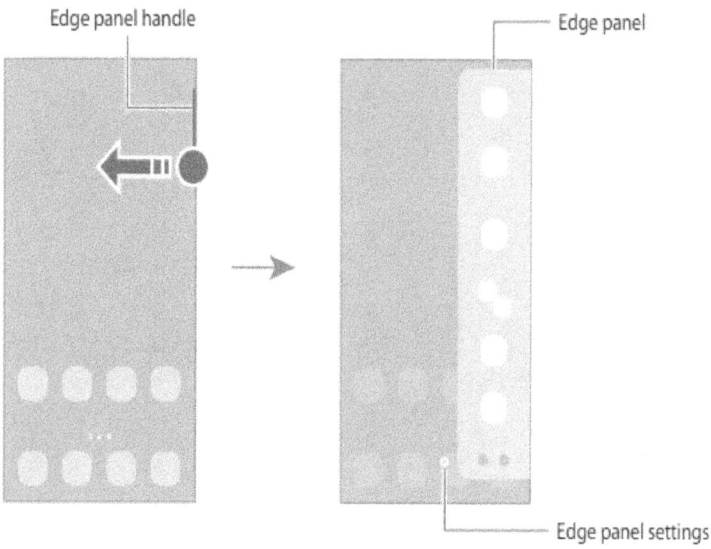

Lock screen

Turn off and lock the screen with the Side key. The screen will also turn off after being inactive for some time depending on your settings. Double tapping the screen can turn it on or off.

How to change the lock screen method

Your lock screen pattern or style can be changed. Set up locks like the Pattern, Password and PIN to protect your personal information, an unlock code will be required of your when your turn on the device.

Note: A factory data reset will take place if your enter the wrong passcode countless times in a row, so endeavor to remember your screen lock method

To change the screen lock:

Launch the Settings application and tap Lock screen then click on Secure lock settings, to enable Automatic factory data reset, unlock your screen with the recent screen lock method and tap the Auto factory data reset switch.

Indicator icons

The icons below are shown on the device status bar, take note of them.

Icon	Meaning
⊘	No signal
▁▃▅	Signal strength
R▁▃▅	Roaming (outside of normal service area)
G↕	GPRS network connected
E↕	EDGE network connected
3G↕	UMTS network connected
H↕	HSDPA network connected
H+↕	HSPA+ network connected
4G↕ / LTE↕	LTE network connected

Icon	Meaning
🛜	Wi-Fi connected
✶	Bluetooth feature activated
📍	Location services being used
📞	Call in progress
☎	Missed call
💬	New text or multimedia message
⏰	Alarm activated
🔇 / 📳	Mute mode / Vibration mode
✈	Airplane mode activated
⚠	Error occurred or caution required
🔋	Battery charging / Battery power level

44

Note:

- Some apps don't allow the status bar to show on them until you swipe downwards on the screen.
- Most icons that are listed above may only appear when you open the notification panel.
- Based on your device model or service provider the icon may appear differently.

Screen capture

You can take an image of the current happenings on your screen with this feature. To capture a screenshot, press and hold the Volume downward and side key together. All screenshots are recorded in the Gallery.

Note: Some applications are not in support of this feature (Screenshot).

Use the image below to customize your screen shot.

- : This can capture more content that are below the screen and beyond.

- ✏️ : To write something on the screenshot or draw on it click this icon.

- ⌇ : Click this icon to share your screenshot with friends.

Note: To allow the toolbar to appear on your screenshot, go to the Settings menu and click on Advanced features and select Screenshot then finally tap the Screenshot toolbar switch to activate it.

Notification panel

Notifications received on your device will appear on the notification panel. The indicator icon for the notification will appear on the notification panel when a new notification is received. Click on the notification to view it details.

From the notification panel, use the function below.

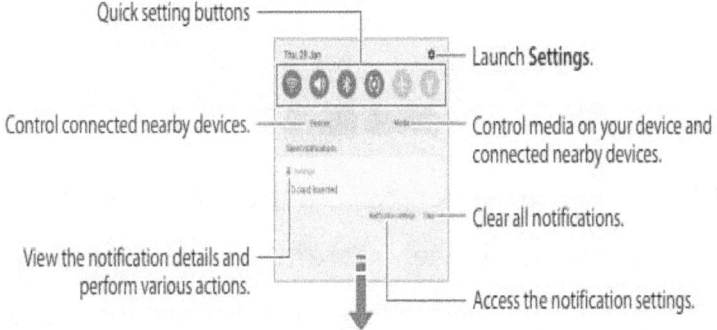

Making use of quick setting buttons

To turn some features on, press the quick settings button. To add more buttons on the notification panel click ⊕ and swipe downwards to view more notifications received.

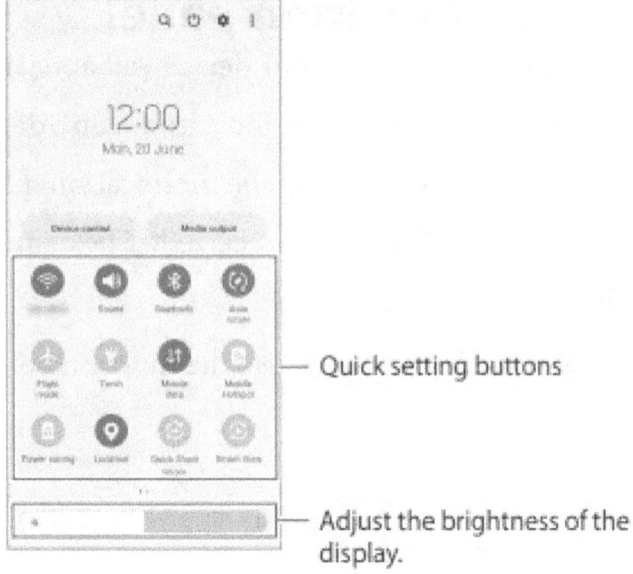

— Quick setting buttons

— Adjust the brightness of the display.

Click the text under each button to change the feature settings. To view more information, press and hold the button. To rearrange the buttons move it to another place after tapping ⁝More options and Edit button.

Controlling media playback

Manage the playback of your videos and music.

1. Go to the gadget notification panel and tap Media.
2. To enable you control media playback, tap the icon that appears on the controller.

Controlling nearby devices

To control nearby devices, follow the steps below.

1. Click on the Device option in the Notification panel.

    Nearby gadgets will appear.

2. To begin controlling, click on any nearby device.

Inputting text
Keyboard layout
Click on the Samsung keyboard to input texts.

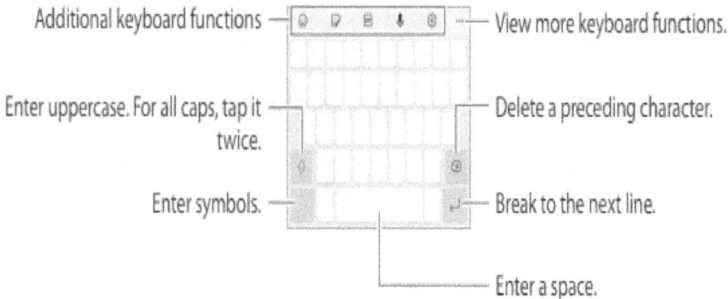

How to change the input language

To change the language of input, click on the ⚙ Settings icon and tap Languages and types then click Manage input languages. To change between languages, swipe to the left or right on the Space bar.

How to change the keyboard

Click on ⠿ in the navigation bar to change the keyboard type.

You can go to the ⚙ Settings app to change the keyboard type by tapping Languages and types then select a language and choose a keyboard type.

Note: Open Settings and click on General management then select the Keyboard list and default then click Keyboard button on navigation in case the ⌨ Keyboard button does not appear.

Additional keyboard functions

- ☺: Click to add emojis to text
- 😀: Tap to add sticker in your conversation
- GIF: Show expressions by sending GIFs
- 🎤: Say a word to input the text.
- ⚙: Tap to change the keyboard settings
- ⋮: Click to view more keyboard features.
- 🔍: Touch to look for apps or chats.
- 🈯: Click to translate texts into another language.
- 🎵: Tap to share your song links.
- ▶: Download interesting movies.
- 📋: Copy and paste texts on your Clipboard.

- ⟨✥⟩ : Edit text with this feature.

- ⌨ : Use this button to change the keyboard layout.

- ⌗ : Click to change the keyboard size.

- ☺ / 😛 : Transfer sticker.

Note: Some features may not be available depending on your gadget's model.

Copying and pasting

1. Hold a text for long.
2. Click Select all or tap ◉ ◉ to highlight text to be copied.
3. The text you have copied will automatically be pasted in your clipboard when you tap Copy or Cut.
4. Click Paste after pressing and holding the location you want to paste the text.

   To paste a text that you have recently copied, enter the clipboard and click on the text.

Application and features

Galaxy Store

Go to the Galaxy Store to download apps that are not available on your device by default. All default apps will appear when you first launch the Apps list.

Go to the Galaxy store. View apps by category or tap the 🔍 Search icon to look for apps.

Note:

- Some apps may not display depending on your service provider.

- to change the settings for auto update, press ☰ Menu then ⚙ Settings and click on Auto update apps and select an option.

Play Store

Go to the Google Play Store application to download apps.

Launch the app. View apps by category or tap the 🔍 Search button to look for an app by keyword.

Note: To change the auto update settings, go to Settings and tap Network preferences then Auto update and select an option.

Uninstalling or disabling apps

Press and hold an app icon and choose an option below.

- **Uninstall**: Take away apps that are downloaded already.

- **Disable**: Default apps cannot be uninstalled but hidden.

Note: All application does not support this feature.

Enabling applications

1. From Settings, click Apps and tap ▼ then Disabled

2. Click OK and press and app then select Enable.

Setting app permissions

Give apps access to your device information by setting permissions for them. Some applications need permission to function properly.

Go to the Settings menu and go to the Apps sections to check app permission. To view the app permission settings, choose an application and press Permission.

To check and change app permission settings do the following:

- Go to Settings and tap Apps, then ⋮ More option and Permission manager.

Note: If apps are not granted permission to your device, you be limited from using some exclusive features of the app.

Chapter Four
Phone application

From the Phone app you can;

- Make audio calls
- Receive calls
- Make video calls
- Decline a call that you don't want to answer.

Making calls

1. Press Keypad in the Phone application.
2. Enter the number that you want to call on the keypad.
3. To place an audio call click 📞 then press 📹 or 📹 to place a video call.

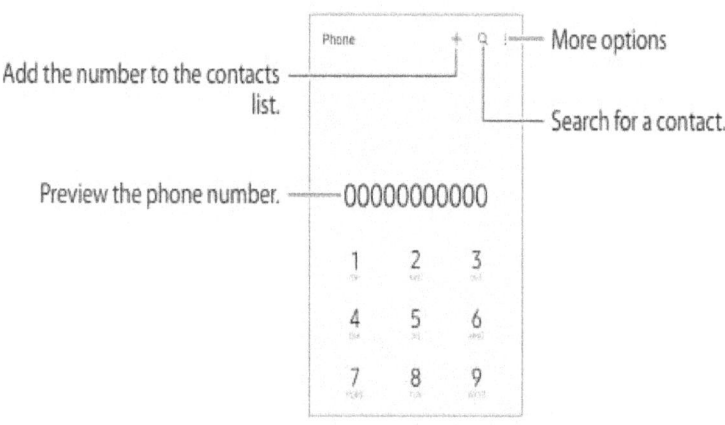

Making calls from call history or contacts list

- From Phone, click on Recent or Contacts and move the contact you want to call to the right.

Tip: Click⋮More Options then Settings and tap Other call settings and finally select the switch for Swipe to call or text to activate the feature if it is disabled.

Making an international call

1. Launch the Phone app and click Keypad.
2. Hold "0" for long until you see "+".
3. Enter the country code and the phone number and click 🕽.

Answering a call

Take 🕽 outside the big circle to take a call.

Rejecting a call

Drag 🕾 outside the big circle to decline a call.

To end a call with a message, Swipe the Send message bar upward and select and message person calling.

To customize a message to send to the caller when rejecting the calls

1. Go to the Phone software and click ⁝ then Settings and tap Quick Decline message
2. Tap Quick decline messages, enter a message and tap ✛.

Blocking phone numbers

1. Open the Phone application and tap ⁝ then Setting and finally click Block numbers.
2. Go to the Recent menu or the Contacts menu and select a contact to block and tap Done.
3. To enter the number that you want to block manually, click Add phone numbers and enter the number and click ✛.

   Tip: Calls coming from blocked numbers will be recorded in your call log as notifications will not get to you.

Note: To block calls that are coming from unidentified callers click "Block unknown/Private numbers".

Options during calls

- **Add call**: You can make another call while on an ongoing call. The call that you are currently on will be put on hold if you make another call. If you stop the second call, the first call will continue automatically.
- **Hold call**: To keep a call on Hold, press "Hold".
- **Bluetooth**: Make a call on a connected Bluetooth device.
- **Speaker**: Keep your phone 4cm away from your ears when using the Speaker.
- **Mute**: Click Mute to turn off the mic for other not to listen to your call.
- **Keypad** / **Hide**: Show or Hide the keypad.
- 📞: Click to end a call.
- **Switch**: Switch between the front and rear camera on a video call.

Note: Depending on your service provider, not all features are available.

Contacts

From this app, you can create a new contact and store it on your phone.

Creating a new contact

1. Go to the Contacts application and tap ✛.
2. Select a storage location to save the contact.
3. Enter the contacts criteria and select Save.

Importing contacts

Bring contacts from other device to your device.

1. From Contacts click ☰ and tap Manage contacts.
2. Select Import or export contacts then Import.
3. To import contacts, follow the screen guidelines.

Syncing contacts with your web accounts

Link all your device contacts with your Samsung account.

1. Go to the Settings application and tap Account and backup

2. Select Manage account and choose and account to sync your contacts with.

3. Choose Sync account and select Contacts to start making full use of the feature.

Searching for contacts

To look for contact in your device

1. Launch contact and tap 🔍 at the upper part of the contacts list and input the search criteria.

2. Choose the number and click one of the following actions:

- To palace a voice call, select 📞.

- Tap 📹 or 📹 to place a video call.

- To compose a message, select 💬.

- To compose and email, tap ✉.

Deleting contacts

1. From Contacts, click ⋮ and tap Delete contact

2. Mark contacts to be deleted and tap Delete.

Tip: Click More options and Delete on contacts to take them away one after the other.

61

Sharing contacts

To distribute your contacts with family and friends do the following:

1. Select ⋮ More options and tap Share contacts in the Contacts application.
2. Look for the number to share and tap Share.
3. Select a sharing method.

Creating groups

Crate a group that friends and family members can be added to.

1. Select ☰ and tap Groups in the Contacts app.
2. Select Create group and follow the onscreen instructions.

Merging duplicate contacts

Combine two or more contacts with the same identity.

1. Click Manage contacts under ☰ in the Contacts application.
2. Tap Merge contacts and choose merge.

Chapter Five
Messages application

From this app, you can view messages, compose and send messages.

Note: Receiving or sending messages when roaming may attract more charges.

Sending messages

1. Launch the ◉ Messages application.
2. Choose a person to send the message to.

   Press and hold ᴵᴵᴵ to record a voice message to the receiver and remove your finger to send the message.

   Note: This button ᴵᴵᴵ will only show when the text input field is empty.
3. Tap ◉ to send the message.

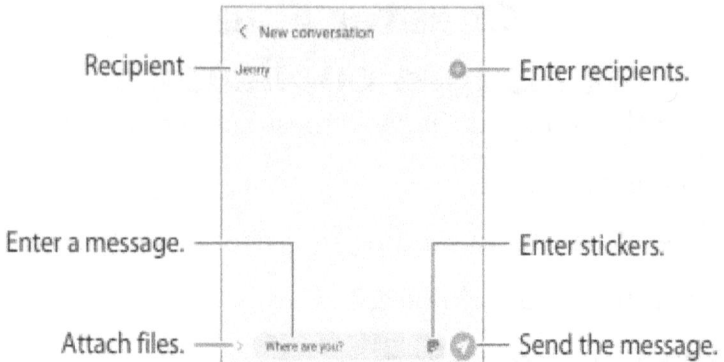

Viewing messages

1. From the 💬 Message application, select the conversation.
2. Select a contact from your message list.
- Spread two of your fingers apart on the screen to increase the size of the text.
- If you want to respond a message, touch on the text input space and then type a message. Click 🔵 after typing your message.

Sorting messages

Messages can be managed easily when they are categorized.

- Go to the 💬 Messages application and select Add category after clicking Conversations.

Tip: To add a category, click ⋮, More options and tap Settings then Conversation categories.

Deleting messages

To remove a message from your device, press and hold it then click on the Delete icon.

Changing message settings

To change the settings for your message application:

- Launch 💬 Messages and tap ⋮ then Settings.

Internet

This application can enable you to search the web and source for more information and also bookmark your favorite pages for easy access.

1. From the "Internet" app, enter your search criteria on the search bar and tap Go.
2. To use the toolbars, swipe downward on the screen.
3. To switch between tabs quickly, swipe to the left or right.

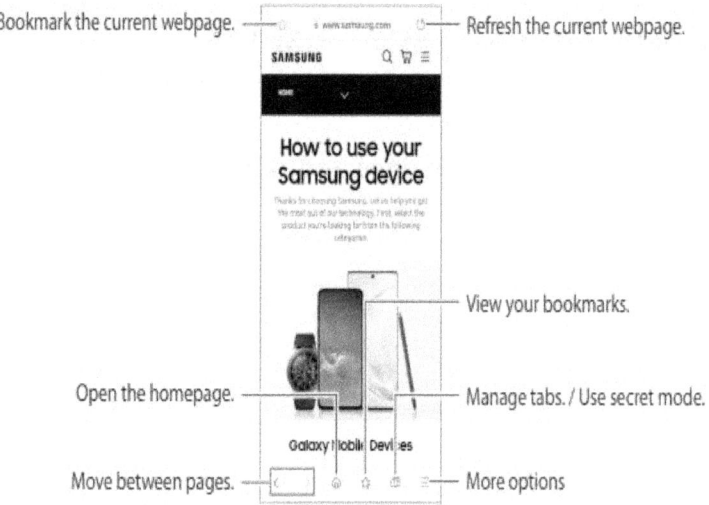

Using secret mode

Use the Secret mode feature to browse on the web to prevent others from viewing your browsing history.

1. Tap on 🗔 Tabs and choose the Turn on Secret mode feature.
2. To start using the Secret mode, click on the Lock Secret mode button and tap Start then enter a password

   The toolbars color will change if the secret mode is enabled. Click on 🗔 Tabs and select Turn off secret mode to deactivate this feature.

Note: You cannot screenshot on the secret mode.

Chapter Six
Camera

The camera is an exclusive application that enables you to capture pictures and record videos.

Camera etiquette

- Require permission from others before taking pictures of them.
- Don't take pictures or record videos in restricted areas.
- Don't take pictures or record videos when people's privacy can be violated.

Taking pictures

- Launch the 📷 Camera application from the Apps list or the Home screen, you can also open the camera app by pressing the side key twice from the lock screen.

Note:

- Not all the features of the camera will open when you access the camera from the lock screen.

- The camera will go off automatically when not in use.

1. Choose where you want the camera to focus on in the picture.

   Pull the brightness adjustment bar that display above or beneath the round frame to adjust the camera brightness.

2. Tap ◯ to capture the image.

   Swipe to the left or right on the camera screen to change between shooting modes.

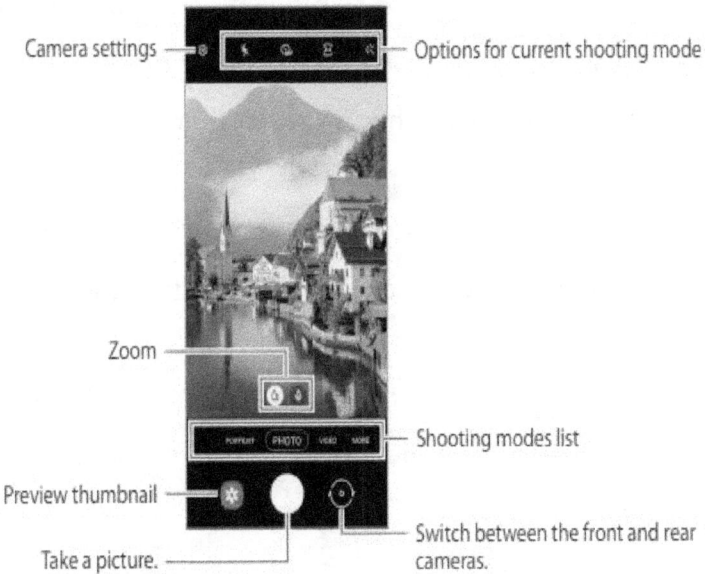

Note:

- Clean the lens of the camera if the picture that was taken appears blurry.
- Your gadget may not function properly on some camera modes that requires a high resolution if the lens of the camera is damaged or dirty.
- The maximum capability for recording a video may vary depending on the resolution.

Using zoom features

To zoom in or out on an image, select ✿ / ✧ or pull it to the left side or right. You can also zoom in by spreading two of your fingers apart on the screen then bring them together on the screen to zoom out.

- ✿ : To snap simple photos or record standard videos, make use of the wide-angle camera.

- ✧ : To capture images or record videos by making the subject wider, use the telephoto feature.

Note: You can only perform the zoom action on the rear camera.

Locking the focus (AF) and exposure (AE)

Lock the area of focus and exposure of the selected area to prevent the camera form adjusting its settings based on the subject or light changes automatically.

The AF/AE frame display and the focus and exposure settings will be locked automatically when the area to focus on is pressed and held down. The settings will not unlock even after taking a picture.

Note: This feature might not be available depending on your shooting mode.

Using the camera button

- To initiate a video recording process, click and hold the camera button.
- To capture a burst (many shots at a time), swipe the camera switch to the edge of the screen and hold it.
- Move the additional camera button to the screen to capture images comfortably. To make

use of the feature effectively, click ⚙ Settings and tap Shooting methods and click Floating shutter button on the preview screen.

Options for current shooting mode

While on the preview screen, you can see the following options:

- ⚡: Turn the flashlight on or off.

- ⏲off: Set a duration for the camera to capture picture automatically.

- 3:4 : Select a photo aspect ratio.

- 8x : Click this button to select a frame rate

- 9:16 : Select this icon to pick an aspect ratio.

- FHD/3D : Select to pick a video resolution.

- ✨: Tap to add beauty effect.

- ○ : On this mode your main focus is the subject that is in the circular frame and blurs the picture that is outside the frame.

- ⊕ : This icon changes the color tone of food in FOOD mode.

Note: Depending on the shooting mode that you are using, the features may differ.

Photo mode

The camera app will adjust the shooting made depending on your environment to give you the best satisfaction when taking a photo.

From the shooting mode screen, select PHOTO and click ⃝ to snap picture.

Taking selfies

You can take pictures of yourself with your device front camera.

1. Swipe up or down on the camera screen or tap the ⊙ Switch icon to use the front camera change to the front camera
2. To capture a selfie, your focus should be on the front camera lens.

   To take a landscape shot or wide-angle shot select 👥.

3. Click ⭕ Capture to take the picture.

Applying filter and beauty effects

Before you take the picture, select a beauty effect you want and adjust the skin quality.

1. Touch ✨ on the preview screen to add effect to the picture.
2. Pick your effect and take the photo.

Video mode

The camera app will adjust the shooting made depending on your environment to give you the best satisfaction when taking record of a video.

1. To initiate the video recording process, click VIDEO and tap 🔴 Record.

   To capture a moment from the video that you are recording, click 📷 Capture.

2. To end the process, click ■ Stop.

Note: The feature "Optical zoom" may not work in an area without sufficient light.

Portrait mode

The portrait mode feature enables the subject to standout clear and the background blurry.

1. From the Shooting modes list, click PORTRAIT MODE.
2. To change the blur level, move the adjustment bar.
3. Select ◯ Capture when you see Effect ready as a pop up on the screen.

Background blur adjustment bar

Note:

- This feature should be used in a place with sufficient light.

- If the following happens, the background blur might not function well:
  - Moving phone and subject.
  - When the subject in focus is so small.
  - Subject and background has the same color.
  - Plain subject.

Pro mode

When you are making changes to different shooting modes like ISO value and exposure value, this feature can enable you take pictures.

o Choose PRO under the MORE menu in the shooting modes list. Select an option and edit its settings. Click ◯ Capture.

Panorama mode

This feature make you take pictures and merge them together as a wide section.

1. Select "PANORAMA" under "MORE" in the Shooting modes list.
2. To take the gadget to one direction, press the ◯ Capture icon.

The frame should be in the viewfinder of the camera. When the preview photo is not inside the guide frame or when you do not move the smartphone, the device will stop taking pictures.

3. Tap ⏹ Stop to end the process.

Note: Avoid the process of taking pictures of empty sky or plain wall.

Food mode

You can get the vivid colors of food with this feature.

1. From the Camera, click on "More" and tap "Food" on the shooting modes list

2. To make a highlight, tap the screen and move the circular frame over to the background. The background that is not inside the circular frame will be blurred.

   To adjust the size of the circular frame, pull the corner of the frame.

3. Change the quality of the color by moving the slider.

4. Touch ◯ Capture to take the picture.

Macro mode

Close range images can be taken with this feature.

- Go to the Shooting modes list and select "More" then "Macro".

Hyperlapse mode

Record, actions like moving vehicles or humans and watch them as fast moving videos.

1. From the Shooting modes list, click "More" and tap "Hyperlapse".
2. Tap this icon and select a frame rate.
3. Click Record to start
4. Click Stop to end recording.

Deco Pic mode

Use this mode to record vides or capture images with stickers or emojis.

- From the list of Shooting modes, click "Deco Pic" under the "More" menu.

Gallery

All the pictures and videos that you have taken with the camera app can be seen in the Gallery application of your device. From the gallery application, you can create folders to manage your gallery app.

Using Gallery

Launch the Gallery application from the Apps list or the Home screen.

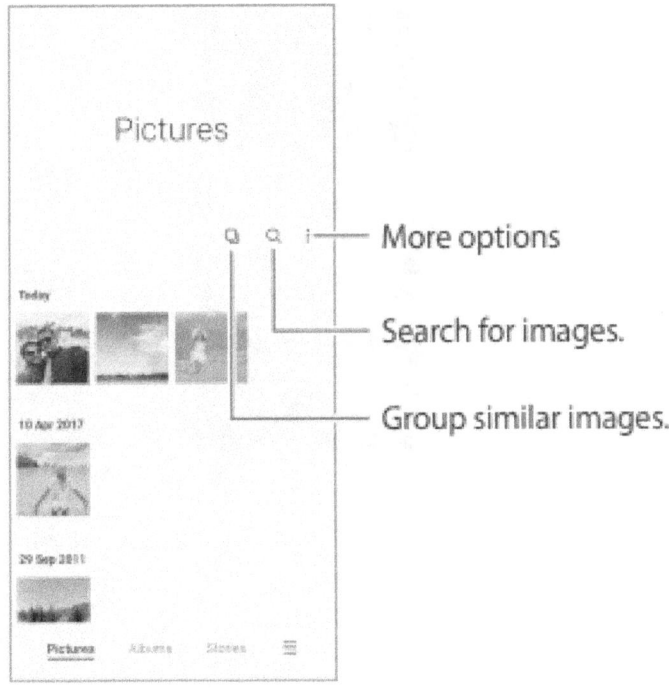

Grouping alike images

- To combine similar images together in the Gallery app, tap ▢. All pictures that are in the group will appear when you click the preview picture.

Viewing images

To view images, open the Gallery app. Swipe the device to the left or right to view other images.

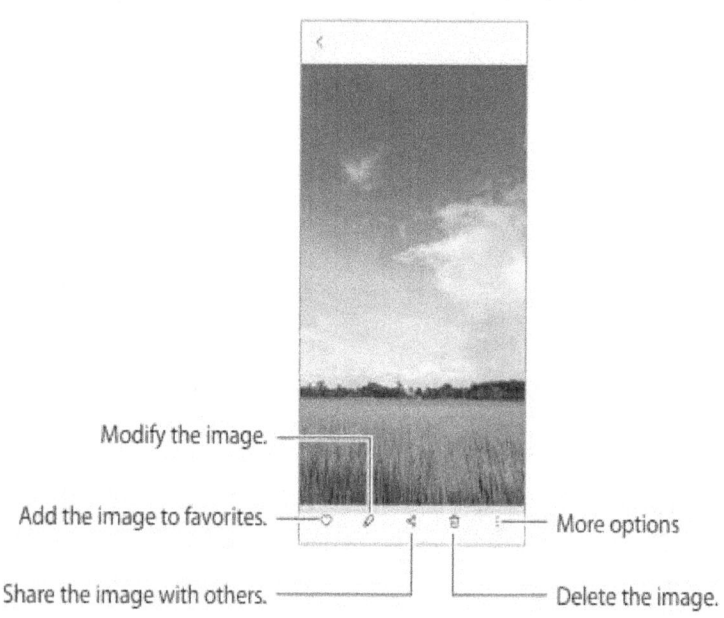

Cropping enlarged images

1. Launch the Galley menu and choose the image that you want to crop.
2. Spread your fingers apart on the part that you wish to save and tap ⬚.

   The cropped part will automatically be saved as a file.

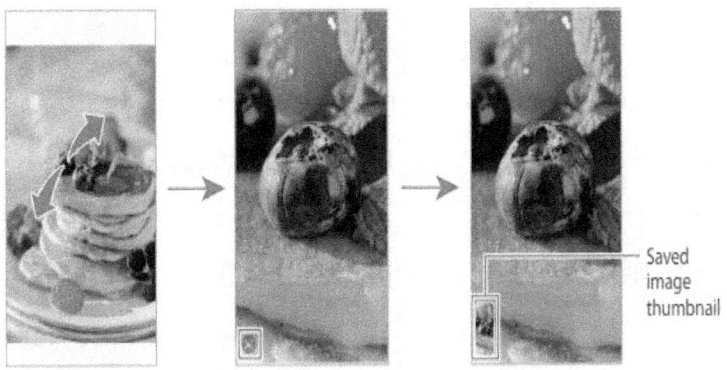

Saved image thumbnail

Viewing videos

To view images, open the Gallery app. Swipe the device to the left or right to view other images.

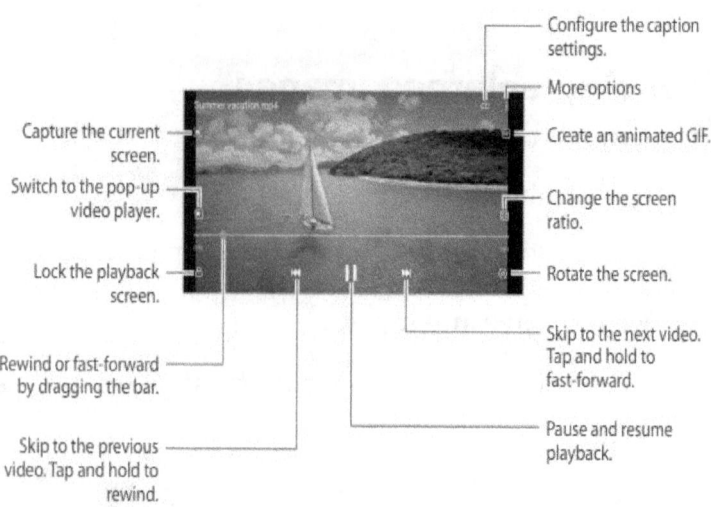

To change the screen brightness, move your fingers upward or downward on the left hand side of the screen. Drag your finger upward on the right hand side of the screen to adjust the volume. Double tap on the left side to rewind and double tap on the right side to fast forward.

Albums

Create folders or albums to organize your videos and pictures in the gallery app for easy access.

1. From Gallery, select Albums and tap ⋮ More and click Create album.

2. Click on the album that you have created and tap Add item to move item to the folder.

Stories

On normal operation, the device read the date and location tags of pictures and videos saved in your device automatically. It also arranges pictures and videos then make stories for you.

- o Go to the Gallery app and click Stories then choose a story.

- Touch a story and tap⋮then click Add or Edit to add or remove pictures.

Using the Trash feature

Deleted pictures are temporarily kept in the trash.

The files in the trash will be removed permanently 31 days after they are deleted.

- o To enable trash, go to the Gallery app and tap ☰ then click Settings and tap Trash.

Tip: To view all files go to the Gallery app and tap ☰ then trash.

Multi window

The Multi window feature is a feature that help you use two apps in the split screen view and multiple apps in the pop-up view.

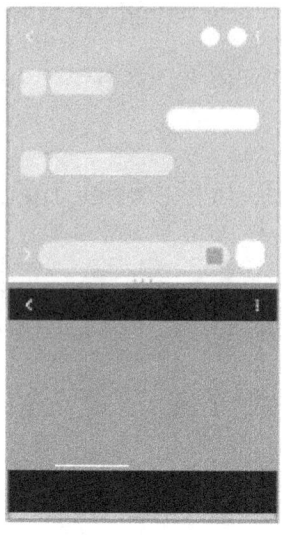
Split screen view

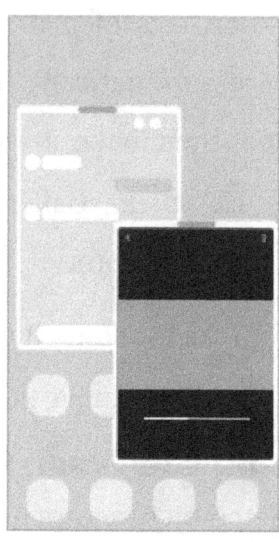
Pop-up view

Note: This feature is not supported by all applications.

Split screen view

1. Click on the Recent button to open the list of apps that have been used recently.
2. Swipe to the left or right, select an app icon and click "Open in split screen view".

3. While you are on the application list, select another app.

Launching apps from the Edge panel

1. To open apps from the Edge panel, take the Edge panel button to the middle of the screen.

2. Hold an app for some time and take it left and leave it where you see "Drop here to open".

   The application you choose will launch in the split screen view.

Note: You can set an app to open in the split screen view when you tap it once. Click on the ✏ Edit icon and tap the ⋮More option button then select "Tap" and click on "Open in split screen view". You can open the recently used app inn the split screen view from the Edge panel if the option "Show recent apps" is selected.

Adding app pairs

Make use of the Edge panel to launch apps that are used often in the split screen view.

1. Touch the circles between apps in the split screen vies to between their windows.
2. Click ⊞ Add.

    The two apps being used in the split screen will be saved as app pairs in the Edge panel.

Adjusting the window size

If you want to alter the size of the application windows, move circles between them.

If you want to maximize app windows, drag the circles between them to the edge of the screen.

Pop-up view

1. Click on the Recent button to see a list of recently used applications.
2. Swipe to the left or right then touch an app icon and select "Open in pop-up view" to open the app in the pop-up view.

The window will be minimized in the pop-up windows when you tap the Home button.

Click the app icon to use the pop-up window again.

Launching apps from the Edge panel

1. Move the handle of the Edge panel to the middle of the screen to launch an app from the panel.
2. Press and hold the app and move it left, then drop it where you see "Drop here for pop-up view".

The selected app will open on the pop-up view.

Samsung Notes

This app is an app that allows you create note of past, present and future events, or save records. Text can be entered in this app either by typing, handwriting or drawing. You can also add pictures, videos, audio and attachments to your notes.

Creating notes

1. Select the ⓔCreate note icon on the Samsung Note app.

   Click ⓐ or ⓑ to change the input method.

   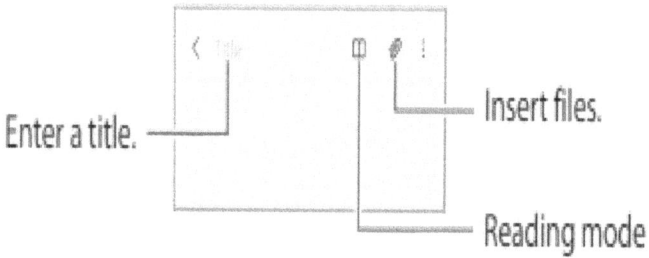

2. Click the Back button to save it, or click the same Back button and tap save as file to save the note in another format.

Deleting notes

Press and hold a note that you want to delete and click "Delete".

Galaxy Shop

Go to the Samsung Galaxy Shot to get info related products and more.

- From the Apps screen launch "Galaxy Shop".

Calendar

You can know the exact date with the calendar app, you can also manage your events by adding future events to your plans.

Creating events

1. From the Calendar app click ⊕ to create events.
2. Input the event information and press the "Save" button.

Syncing events with your accounts

1. From "Settings", click "Account and Backup"
2. Select "Manage account" and choose the account to sync with.
3. Click Sync account and click Calendar to activate this feature.

Enter **Calendar** and select ☰ > ⚙ > **Add account** to place account to sync with. Login with the account that you choose to sync with. If you added the account, a blue circle will display next to the account name.

Reminder

Set your reminder to notify you on your next activity.

Note:

- Make sure that you have an internet connection, to receive the accurate notification.
- Turn on GPS to make use of location. Depending on your service provider, location reminders may not be available.

Starting Reminder

o From Calendar tap ☰ > **Reminder**. 🔔 Will automatically be added to the Apps screen.

Creating reminders

1. From 🔔 Reminder click ➕ Add.
2. Type a word and tap "Save".

Finishing reminders

- From 🔔 Reminder, click ⭕ or select a reminder tap **Complete**.

Restoring reminders

To restore reminders

1. Launch the 🔔Reminder app and select ≡ then **Completed**.
2. Select a group and tap Edit.
3. Look for the reminder that you want to restore and tap "Restore" after seeing it.

Deleting reminders

- Look for a single reminder to delete, click on it and tap "Delete"
- Mark multiple reminder to be deleted and tap "Delete".

Radio

Launch the FM app in the Apps screen.

An earphone which operates as an antenna must be connected to your device before using this app.

The radio app search for available channels and save them so you can easily access them next time.

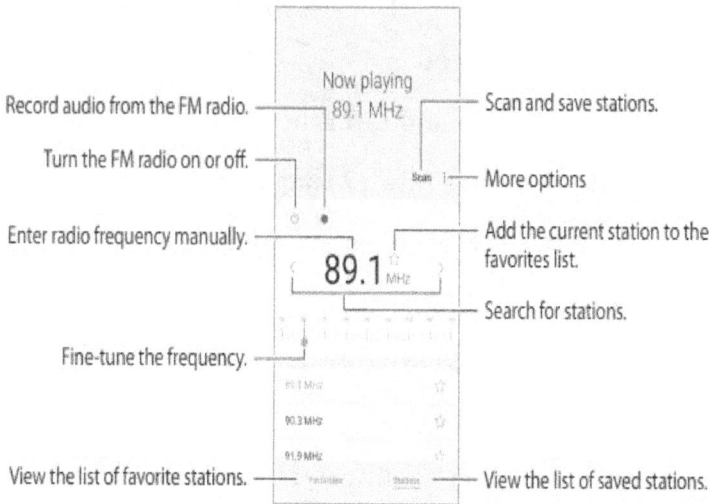

Note: This app may be unavailable, depending on the service provider or the device model.

Playing through the speaker

- Click on ⋮More in the Radio menu and tap "Play through speaker".

My Files

This is an app that enables you view your device media and data and also enables you manage them.

1. From the Apps screen or the Home screen, launch the My Files app.
2. Tap "Analyze storage" to see file that are taking up you space.
3. Click 🔍 Search to look for files in the app.

Clock

This app can make you create alarms, create a countdown on the stopwatch, keep track of an event and check the time in some cities.

- Launch the app (Clock app) from the Apps screen.

Calculator

You can carry out simple and complex calculations with this app.

- o Enter the app (Calculator app) from the Apps screen.
  - ■ ⓛ : This icon is to view the history of your calculations. Select "Clear history" to clear the calculation history. Click 🖩 to exit the history menu
  - ■ You can use the conversion of unit tool by tapping 📏. Units like Length, Area, Temperature can be converted into other units.
  - ■ Tap 🧮 to make good use of the scientific calculator.

Sharing content

There are different methods of sharing content. The example below directs you on how to share images.

1. Select the image or videos to be share from the Gallery application.

2. Click ⁰⁄₀ Share and choose a sharing method.

Note: Extra charges may apply, when files are shared through mobile networks.

Quickly Share content with nearby devices

Use Wi-Fi Direct, Bluetooth or SmartThings to share content with nearby devices.

1. Launch the app (Gallery) and select the image that you want to share.

2. From the other device, enter the notification panel and tap ⊚ Quick share. Select ⊕ Add and drag the switch to add Quick share.

3. Click ⁰⁄₀ Share and then ⊚ Quick share and choose the device to share the image with.

4. Accept the sharing request from the other device.

Google apps

Google apps offers exclusive content and social network. To use this feature, a Google account is required.

- **Chrome**: Go to Chrome to look for information on the web.
- **Gmail**: Send and receive emails on this app.
- **Maps**: View location details for different places around you with this app.
- **Google Play Movies & TV**: Videos like TV programs can be bought lease from this app.
- **Drive**: All your files can be stored in this app's cloud, so that you can access it on any device.
- **YouTube**: View videos of entertainment on YouTube and also share with others.
- **Photos**: View and edit pictures that are stored on your device from the app.
- **Google**: Search for things about sports, news and more.

- **Duo**: This app enables you to make video calls easily.

Note: Some of the apps listed above might be unavailable, depending on your device model.

Chapter Seven
Settings

Alter the settings for your Samsung A05.

1. Go to the quick panel or Apps screen to access the Settings application

2. Tap 🔍 Search to look for settings easily.

Samsung account

- From Settings, click Samsung account and login to your account, or create one if you don't have.

Connections
Options

Connection settings like the Bluetooth, Wi-Fi and other connection settings can be altered from this menu.

- From Settings, click on Connections.
  - **Wi-Fi**: To connect to a Wi-Fi network, turn it on from the quick panel.
  - **Bluetooth**: Send files to other device that supports the Bluetooth feature.

- **Airplane mode**: All wireless connectivity except the Wi-Fi will be turn off when you activate this.

  Warning: Turn on this feature, while on an aircraft or ship to avoid violating the authority.

- **SIM card manager** (dual SIM models): Insert the dual SIM cards into the device and change the SIM card settings to your taste.

- **Mobile Hotspot and Tethering**: Turn on Mobile Hotspot to share your data connection with other device.

Note: Additional charges may apply for using this feature.

Wi-Fi

Turn this feature on to connect to other device connection.

Connecting to a Wi-Fi network

1. Click "Connection" from "Settings" and tap "Wi-Fi" then turn on the feature with the switch that is available.

2. Select a network to connect to from the list. A password is required for network s that carries a lock icon.

Note:

- If a network that you have connected is in range, your device will connect to it automatically with needing a password. Click ⚙ next to the network that you have connected before and tap Auto reconnect then turn it off, if you don't want your device to connect to it automatically.

- Restart the device wireless router or Wi-Fi connection, if you are having issues with connection.

Wi-Fi Direct

With this feature, you can connect to devices via Wi-Fi without needing access point.

1. From Settings, click on Connection and tap Wi-Fi.
2. Tap ⋮ More options then Wi-Fi Direct.

    The available devices will appear.
3. Tap the device that you want to connect to.

    There will be an automatic connection if the other device that you want to connect to agrees to the connection request.

    Tap the device and click "Disconnect" to end the connection.

Bluetooth

You can share files to nearby device with this feature.

Caution:

- Warranty does not cover damages caused to the device by sent or received data via Bluetooth.

- Apps that are not approved by Bluetooth SIG may not be suitable for your gadget.

- Don't use this feature for illegal purposes like pirating and others. Samsung will not be held responsible for the danger of unlawful usage of the Bluetooth feature.

Pairing with other Bluetooth devices

1. Click Connections and tap Bluetooth from Settings, then tap the switch to activate it. Available device will appear.

2. Tap a device to connect with from the list.

   Put your device into the pairing mode if the gadget you want to connect to is not in the list.

   Note: Once you activate Bluetooth, your device will be visible to others.

3. Accept the Bluetooth connection request that is sent to the other device.

Both devices will pair automatically when the connection request has been accepted.

Click the ⚙ Settings icon close to the Bluetooth name and tap Unpair to unpair them.

Sending and receiving data

Share data such as your device contacts and media with other devices that supports Bluetooth, some apps support sharing through Bluetooth. Do the following to share pictures and other media files to other devices.

1. From the Gallery app select a photo that you want to share through Bluetooth.

2. Click ⊰ Share and tap "Bluetooth" then select the device from the list.

Turn on the Bluetooth visibility feature if the device that you want to connect with is not in the list.

3. Tap "Accept" to agree to the connection request.

Data saver

Manage and control your data usage by limiting some application from sending and receiving data on background.

- Click "Connections" and tap "Data usage" and click "Data saver" and turn on the feature all from the Setting app.

Tip: If the Data saver feature is on, this (⟳) will appear on the status bar.

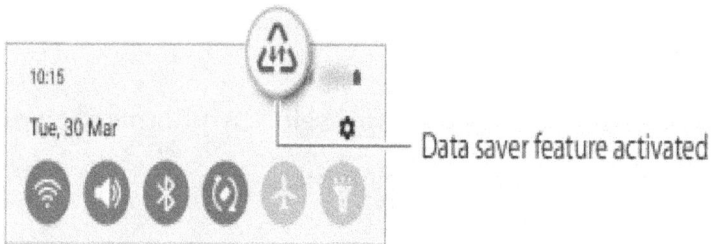
Data saver feature activated

Note: You can enable some apps to use data without restrictions. After tapping "Allow to use data while data saver is on" choose the app you want.

Mobile data only apps

Some apps can be allowed to use data even when you are connected to a Wi-Fi network.

You can set your device to use only data for apps that you want to keep safe. The app will open using the device mobile data, even when the Wi-Fi is turned on.

- Select "Connections" in the "Settings" app and tap "Data usage" then select "Mobile data only apps" and then turn on the data saver. Turn on all the apps that you want to use mobile data for.

Note: Additional charges may apply, using this feature.

Mobile Hotspot

With the Mobile Hotspot, you can share your device data connection with other devices.

1. Tap on "Connections" under the "Settings app" then tap "Mobile hotspot and Tethering" then click Mobile Hotspot.

2. Turn on the feature with the switch.

   If the Hotspot is on, this (📶) will appear on the status bar.

   To change the Hotspot security level and password, tap on "Configure"

3. Activate Wi-Fi on the device that you want to share your smartphone connection with and hit on your Hotspot name. You can also scan the QR code to connect it.

Note: If the Mobile Hotspot is hidden on your device, click "Configure" and tap "Advanced" then "Hidden network" and click the switch to turn it on.

Printing

You can connect your gadget to a printer, using the Wi-Fi or Wi-Fi Direct feature. On your gadget, configure and add printer plug-ins installed. Pictures and other documents can be printed from your devices

Note: Your gadget may not be compatible with some printers.

Adding printer plug-ins

1. From Settings, tap Connection and click More connection settings.
2. Select Printer and click Download plugin.
3. Find a printer plug-ins sand install it on your gadget.
4. Select the printer that is installed.

    Your gadget will look for printers that are connected to the same Wi-Fi network automatically.
5. Look for a printer to set.

Note: Click ⋮ More options and tap "Add printers" to add printers manually.

Printing document

While viewing the document or picture that you want to print, go to the option list and tap on Print then ▼ and tap "All printers" and select the picture or document that you want to print.

Note: The file type may cause the printing method to vary.

Sounds and vibration
Options
Sound types can be changed on your device.

- o From "Settings" click on "Sound and vibration".
- **Sound mode**: Your device can be set to, Ring, Vibrate or put to Silent.
- **Vibrate while ringing**: When you have an incoming call, you can set your mobile phone to vibrate.
- **Temporary mute**: Your device can be put to silent for some period of time.
- **Ringtone**: You can customize a ringtone for your incoming calls if you don't like the default sound.
- **Notification sound**: You can customize a notification sound or use the preset one.
- **Volume**: You can reduce or increase the device sound level.
- **Call vibration pattern**: Set a pattern of vibration pattern for all calls.

- **Notification vibration pattern**: Set a pattern of vibration for notifications.
- **Vibration intensity**: You can increase or reduce the strength of the device vibration.
- **System sound/vibration control**: Change your device settings for all tap to make a sound and vibrate.
- **Sound quality and effects**: Change the Phone sound effect and it quality to your taste.

Note: Some features may be absent, depending on the model of your device.

Notifications

Change the settings of your notifications.

- From Setting, click "Notifications"
- **Notification pop-up style**: Select a pop-up style for notifications and change their settings.
- **Recently sent**: See apps that have sent you notifications recently and change their settings. Select More and tap ▼ then select All and select app from the apps list to change the notification setting for apps.
- **Do not disturb**: Turn this mode on to avoid calls coming from other to emit a sound an also silent notification sound. You can set exception while device is in Do Not Disturb mode. You can also customize a schedule for this feature to be turn on automatically
- **Advanced settings**: Go to the Setting app and set advanced notification setting.

Display

Customize the appearance of your device Home screen or Lock screen.

- o Click Display from Settings.

  - **Light / Dark**: Turn this feature on to use it or turn if off to switch back to normal mode.

  - **Dark mode settings**: Dark mode helps to reduce eye strain while at night.

    Note: While in some apps, this feature may not function.

  - **Brightness**: The device brightness can be adjusted to either increase or decrease it.

  - **Adaptive brightness**: This feature help to adjust your screen brightness based on the lighting conditions of your environment.

  - **Eye comfort shield**: Turn this feature on while you are in a dark environ. It also reduces eye strain and makes you sleep comfortably at night. You can still set a schedule to turn on this feature automatically.

- **Font size and style**: Change the Settings of your font in terms of style and size.
- **Screen zoom**: From here you can increase or decrease the size of things appearing on the screen of your phone.
- **Full screen apps**: Select apps that should be used in full screen aspect ratio.
- **Screen timeout**: Customize how long your device will stay on when unused for some time.
- **Edge panels**: Change the Edge panel settings to your taste.
- **Navigation bar**: change the navigation bar settings.
- **Accidental touch protection**: Configure your device not to detect accidental taps while it is your bag or pocket.
- **Screen saver**: Set your phone to display a screen saver while charging.

Note: Some features are unavailable depending on the model of your device.

Wallpaper

Change the Wallpaper settings for your Home and Lock screen.

- o Select the "Wallpaper" option in the Settings page.

Themes

Go to Galaxy themes and get more themes to change the appearance of icons and widgets on the Home and Apps screen.

- o Select the option "Themes" in the Settings menu.

Home screen

Change the appearance of your Home screen.

- o Select the Home screen option in the Settings menu.

Lock screen

Change the appearance of your Lock screen.

- o Select Lock screen in the Settings menu.
  - **Screen lock type**: Change the default screen lock type that came along with the device.

- **Smart Lock**: Your phone can be set to lock automatically when a trusted location is detected.

- **Secure lock settings**: Change the settings of your screen lock.

- **Wallpaper services**: Allow your device to use Wallpapers like the Dynamic Wallpapers.

- **Clock style**: Change the clock style and color.

- **Roaming clock**: While roaming, allow the clock to show both local and home time zone.

- **Widgets**: Add widget to customize the appearance of the Home screen.

- **Contact information**: Set your phone to show contact details like the Email address and others on the lock screen.

- **Notifications**: Change the display of notification on your device.

- **Shortcuts**: Select the app shortcut to appear on the Lock screen.

- **About Lock screen**: Check some rightful info, like the software version and the lock screen version.

Note: The options above may vary depending on the type of the screen lock method you choose.

Smart Lock

Your device will unlock automatically, when it detects a trusted location, once this feature is on.

When your house or office is set as a trusted location, your phone will unlock automatically, when you are in the house or at the office. Set your Home and Office as a trusted location for security reasons.

- o Select "Lock Screen" in the "Settings app" and click "Smart Lock" and follow the rules on the screen to finish the setup.

Note: This feature can only be used when there a screen lock.

Chapter Eight
Biometrics and security

Change the security settings of your gadget.

- From Settings, click on "Biometric and security"

  - **Face recognition**: Your phone will unlock automatically when you look towards it, if you set face recognition and use your face.

  - **Fingerprints**: When you set your fingerprint on the device, the device will unlock automatically once you place your finger on the fingerprint sensor.

  - **More biometrics settings**: Change the settings for your biometric data.

  - **Google Play Protect**: This feature of your device helps to check for harmful apps and tell you concerning the damage that it may cause.

  - **Security update**: Check the security version of your device regularly and update if necessary.

  - **Google Play system update**: Check Google Play Store and update apps when necessary.

- **Find My Mobile**: When your phone gets stolen or missing, visit (findmymobile.samsung.com) to trace the phone.
- **Private Share**: Share your files privately and securely with the Block-chain technology.
- **Install unknown apps**: Grant permission to your device to install apps from unknown source.
- **Encrypt SD card**: Configure your device to encrypt (encode) all files stored in an SD card.

  Caution: The smartphone will be unable to detect your files that are encrypted. When you reset your phone to factory default settings
- **Other security settings**: Change other security settings for your device.

Note: Some features may be unavailable based on the device's model.

Face recognition

Enroll your face to use the face recognition process to unlock your phone.

Note:

- Your device cannot be unlocked with your face the first time of turning it on, if you use face recognition as a screen lock. To unlock the device after turning it on, enter the PIN, Password or draw the Pattern to unlock your device depending on your screen lock types.

  Note: Once the "None" or "Swipe" security feature is used on your device, all biometric data will be deleted. You will have to register your face again for you to be able to use it.

  Note: This screen lock method has a very low level of security as you device can be unlocked by some of same resemblance with you.

For better face recognition

- If you are on glasses, makeups, masks, hats, the face recognition may not work well.

- When enrolling your face, make sure you do it in an area with sufficient light.
- Capture a very clear image.
- Avoid direct sunlight.

Registering your face

1. Click on Biometric and security in the Settings menu and tap Face recognition.
2. Click "Continue".
3. Select a screen lock type.
4. Make the camera lens clean and look towards the frame of the camera.

Note: If you're having difficulties unlocking your device with your face, hit "Remove face data" to erase your previous registered face and re-enroll it again.

Unlocking the screen with your face

Use your face to unlock your device rather than using the, Password, PIN or Pattern.

1. Go to the "Biometric and security" menu in the Settings app and click Face recognition.
2. Input the screen lock to unlock the screen.
3. To turn on this feature, click "Face unlock".

4. Look toward the camera to unlock your device.

Note: Enter your screen lock in a dark area if the camera could not detect your face.

Deleting the registered face data

1. Click on "Biometric and security" in the Settings app and select Face recognition.
2. Enter your screen lock to open the phone.
3. Select "Remove face data" and tap "Remove"

Note: Face recognition will be removed from all apps and screen lock method, if your turn off the feature.

Fingerprint recognition

Register your fingerprint and store it data on your device for easy and faster recognition.

Note the following:

- The feature may not be present on your device depending on the model.
- With this feature, you can improve your device security. It is also safer than the face recognition.
- Your device cannot be unlocked with your fingerprint the first time of turning it on, if you

use your fingerprint as a screen lock. To unlock the device after turning it on, enter the PIN, Password or draw the Pattern to unlock your device depending on your screen lock types.
- Enroll your fingerprint again is it is not working properly.
- Once the "None" or "Swipe" security feature is used on your device, all biometric data will be deleted. You will have to register your face again for you to be able to use it.

For better fingerprint recognition
- A finger with wrinkles will not be detected.
- Tiny fingers might not be detected.
- Don't take metallic objects close to the sensor, else the feature may malfunction or get damaged.
- Ensure that you clean and dry up the fingerprint sensor.
- Place your finger on the sensor properly.

Registering fingerprints

1. Select "Biometric and security" in the Settings app and click "Fingerprints".
2. Click "Continue".
3. Select a screen lock method to use.
4. Place your finger on the side key and remove it once the device vibrates, do this until the process is complete.

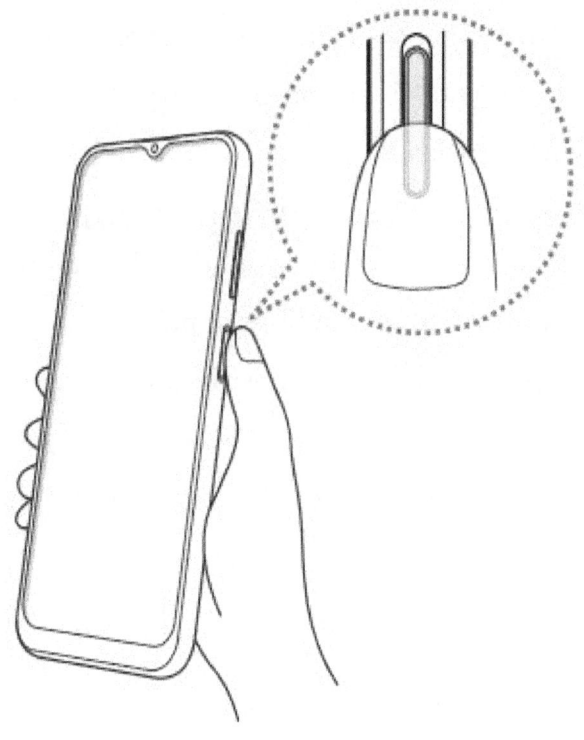

5. After enrolling the fingerprint, click "Done".

   Tip: Tap "Check added fingerprint" to know if your finger has been registered successfully.

Unlocking the screen with your fingerprints

1. Click "Biometric and security" and tap "Fingerprint" from Settings.
2. Use the screen lock pattern that you have set to open the device.
3. To activate the feature, hit "Fingerprint unlock".
4. Place your finger on the sensor to unlock your device with it.

Deleting registered fingerprints

1. Click "Fingerprints" under "Biometric and security" in the Settings application.
2. Unlock your phone with the current screen lock.
3. Tap "Remove" after clicking on the fingerprint that you want to remove.

www.ingramcontent.com/pod-product-compliance
Lightning Source LLC
Chambersburg PA
CBHW070146230526
45471CB00002B/540